I0446614

A DIGEST ON GENETICS

'Genetics'

By

Pyura

Pen Name: **P Y U R A**

A N S H U M A N

2025 Edition

Pyura Books & Research

www.pyurabooks.com

A Digest on Genetics. Copyright © 2025 by Pyura Anshuman. All rights reserved. First published by Amazon in 2023. New edition published in 2025. No part of this book may be used or reproduced, stored in or introduced into a retrieval system, or transmitted, in any form or in any manner whatsoever (electronic, mechanical, photocopying, recording or otherwise), without prior written permission of the copyright owner except in the case of brief quotations embodied in critical articles and reviews.

Pyura Books & Research may be purchased for educational, business or sales use. For information and communication write to publish@machaanlabs.com

Book cover designed by Sadia Akhtar

ISBN   979-8-86-542151-1

∞∞∞∞∞∞∞∞∞∞∞∞∞∞∞∞∞∞∞∞∞∞∞∞∞∞∞∞∞∞∞∞∞∞∞

WHERE NATURE, SCIENCE AND SOUL COLLIDE. 📚

PYURA

Within every strand of DNA lies not just the code of life, but the story of who we are, where we come from, and the endless possibilities of what we can become.

TABLE OF CONTENT

GENETICS

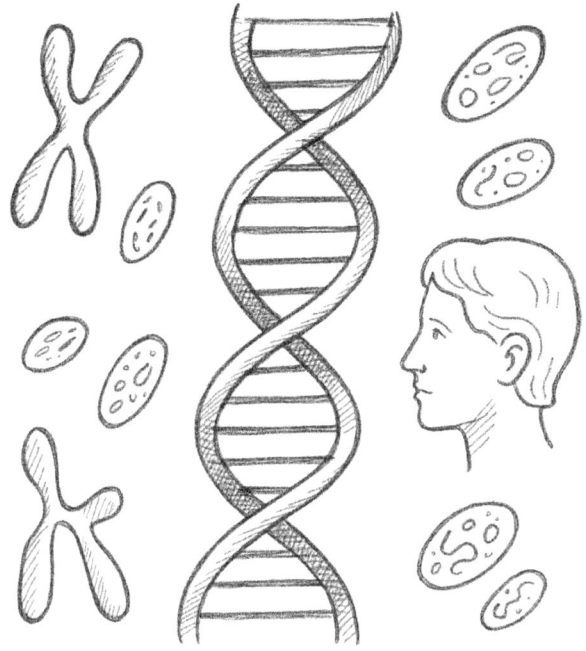

PYURA BOOKS
& RESEARCH

INTRODUCTION TO GENETICS

Genetics as the Study of Genes, Heredity, and Variation

Genetics is a multifaceted scientific discipline that revolves around the study of genes, heredity, and variation in living organisms. At its core, genetics delves into the fundamental units of heredity, known as genes, which carry the genetic instructions necessary for the development, functioning, and inheritance of traits in individuals and populations. Through the exploration of genes, geneticists uncover the underlying mechanisms that govern the transfer of genetic information from one generation to the next, shedding light on the inheritance patterns and diversity observed within species. By understanding the processes of replication, recombination, and mutation, genetics provides invaluable insights into how organisms evolve and adapt to their environments, playing a crucial role in fields as diverse as medicine, agriculture, and evolutionary biology. This study of genes and the interplay between heredity and variation is essential for unraveling the complexities of life and understanding the genetic basis of both health and disease.

Historical context of genetics, from Gregor Mendel's experiments to the discovery of the DNA double helix by Watson and Crick

The historical context of genetics is a captivating journey that spans centuries of scientific inquiry and remarkable discoveries. It all began with the pioneering work of Gregor Mendel in the mid-19th century, whose experiments with pea plants laid the foundation for modern genetics. Mendel's meticulous observations on inheritance patterns revealed the existence of discrete hereditary units, later known as genes, and the principles of dominant and recessive traits. However, Mendel's groundbreaking work remained largely unnoticed until the early 20th century when scientists like Thomas Hunt Morgan began connecting Mendelian genetics with the concept of chromosomes, elucidating the physical basis of heredity.

In the 1940s, the study of genetics advanced significantly with the discovery of DNA as the molecule carrying genetic information. One of the pivotal moments in this journey was the discovery of the DNA double helix structure by James Watson and Francis Crick in 1953. Building on the work of Rosalind Franklin, they proposed a model that elegantly explained how DNA's structure allowed for its replication and the transmission of genetic information during cell division. This groundbreaking discovery marked the birth of molecular genetics, revolutionizing our understanding of heredity and opening up a new era of genetics research.

The fusion of Mendelian genetics with the discovery of the DNA double helix formed the basis for modern genetics, laying the groundwork for numerous breakthroughs and applications in fields such as biotechnology, medicine, and genetic engineering. The historical context of genetics stands as a testament to the relentless pursuit of knowledge and the immense impact that scientific curiosity and innovation can have on unraveling the mysteries of life.

Detailed structures of DNA, including nucleotides, base pairing, and the concept of antiparallel strands.

DNA, or deoxyribonucleic acid, is a remarkable macromolecule that serves as the blueprint of life, encoding the genetic information essential for the development, functioning, and reproduction of all living organisms. At its core, DNA is composed of repeating units called nucleotides, each comprising a sugar molecule (deoxyribose), a phosphate group, and one of four nitrogenous bases: adenine (A), thymine (T), guanine (G), and cytosine (C). The genius of DNA lies in its base pairing rules: adenine pairs with thymine (A-T) through two hydrogen bonds, and guanine pairs with cytosine (G-C) through three hydrogen bonds. This complementary base pairing enables the formation of the iconic DNA double helix, wherein two strands wind around each other in a twisted ladder-like structure. Moreover, these two strands run in opposite directions, a phenomenon known as antiparallel orientation, where the 5' end of one strand aligns with the 3' end of the other. The detailed structure of DNA and its specific base pairing mechanisms not only facilitate precise DNA replication during cell division but also underpins the essential processes of transcription and translation, making DNA the fundamental molecule of genetics and the cornerstone of life's diversity and complexity.

DNA STRUCTURE AND REPLICATION

Process of DNA replication, involving enzymes like DNA helicase, DNA polymerase, and DNA ligase

D NA replication is a crucial cellular process that ensures the faithful duplication of the genetic information stored within DNA molecules. It occurs during the S phase of the cell cycle, preparing cells for division. The process involves a series of coordinated steps carried out by specialized enzymes. Initially, the DNA double helix unwinds at specific origins of replication, facilitated by an enzyme called DNA helicase. As the strands separate, single-strand binding proteins stabilize them, preventing reannealing. To synthesize new DNA strands, DNA polymerase comes into play. This enzyme adds nucleotides to the growing strand, following the rules of complementary base pairing. The leading strand is synthesized continuously in the 5' to 3' direction, while the lagging strand is synthesized in short fragments called Okazaki fragments. DNA ligase then joins these fragments to create a continuous strand. The process proceeds bidirectionally from the origin until the entire DNA molecule is replicated. The faithful replication of DNA is essential for passing accurate genetic information to daughter cells during cell division and for maintaining the genetic stability crucial for the functioning and survival of organisms.

The Significance of DNA Replication in Cell Division and Inheritance

The significance of DNA replication in cell division and inheritance cannot be overstated as it serves as a foundational process for the perpetuation of life and the transmission of genetic traits from one generation to the next. During cell division, DNA replication ensures that each daughter cell receives an identical copy of the genetic material present in the parent cell. This faithful replication is vital to maintain the genetic integrity of the organism and ensures the proper functioning and development of cells and tissues. In sexual reproduction, DNA replication plays a pivotal role in the production of gametes (sperm and egg cells), where genetic information from each parent is combined to create a unique genetic makeup in the offspring. Additionally, DNA replication enables the inheritance of traits and characteristics across generations, allowing for the preservation and evolution of species over time. Errors in DNA replication, known as mutations, can lead to genetic variations that can be advantageous, neutral, or harmful, contributing to the diversity and adaptability of populations. Thus, DNA replication forms the cornerstone of cell division and inheritance, enabling the continuity of life and shaping the rich tapestry of genetic diversity observed in the natural world.

GENES AND CHROMOSOMES

Genes and alleles, explaining how they determine specific traits

Genes are fundamental units of heredity found in the DNA of living organisms. They act as blueprints for the production of specific proteins or functional RNA molecules, which in turn carry out various cellular processes and determine an organism's traits. Each gene occupies a specific location on a chromosome and is composed of a unique sequence of nucleotides that code for a particular product. Alleles, on the other hand, are alternative forms or variants of a gene that can occupy the same gene locus on homologous chromosomes. Alleles may have slight differences in their DNA sequence, resulting in distinct variations of the trait they influence. In diploid organisms, such as humans, an individual inherits two alleles for each gene, one from each parent. The combination of alleles that an individual carries contributes to the specific observable traits or characteristics they exhibit. Some traits are determined by a single gene with two distinct alleles (e.g., hair color), while others are influenced by multiple genes and environmental factors (e.g., height). The interplay between genes and alleles is at the heart of the intricate process of inheritance and the vast array of traits observed in the rich tapestry of life.

Gene expression, including transcription and translation, and the role of regulatory elements like promoters and enhancers

Gene expression is a complex and tightly regulated process that involves the conversion of genetic information stored in DNA into functional products such as proteins or non-coding RNAs. It encompasses two main stages: transcription and translation. During transcription, a specific gene's DNA sequence is copied into a complementary RNA molecule by an enzyme called RNA polymerase. The process begins when regulatory elements like promoters and enhancers, located near the gene, recruit transcription factors that initiate the binding of RNA polymerase to the gene's promoter region. Enhancers can act over long distances, looping the DNA to facilitate interactions between regulatory elements and the transcription machinery. Once transcription is complete, the newly synthesized RNA molecule, known as messenger RNA (mRNA), undergoes processing, including the removal of introns and addition of a protective cap and tail. Subsequently, during translation, the mature mRNA travels to the ribosomes in the cytoplasm, where the genetic code is translated into a sequence of amino acids to form a polypeptide chain. The coordination and fine-tuning of gene expression are critical for ensuring that the right genes are activated at the right time and in the right cells, playing a pivotal role in the development, differentiation, and proper functioning of an organism. Regulatory elements like promoters and enhancers serve as key players in this intricate process, modulating gene expression and contributing to the remarkable complexity and diversity of life.

Structure and function of chromosomes, covering homologous pairs, sex chromosomes, and the cell cycle

Chromosomes are thread-like structures composed of DNA and proteins that carry an organism's genetic information. They are essential for the organization, protection, and inheritance of genetic material. In diploid organisms, chromosomes are found in pairs, with one member inherited from each parent, forming homologous pairs. These homologous chromosomes contain similar genes, although they may have different alleles, contributing to the diversity of traits within a species. Among the homologous pairs, the sex chromosomes determine an individual's sex; in humans, females have two X chromosomes (XX), and males have one X and one Y chromosome (XY). During the cell cycle, chromosomes undergo a series of events, including DNA replication, condensation, alignment, and segregation, ensuring that each daughter cell receives the correct number of chromosomes during cell division. This process, along with the precise distribution of genetic material, is crucial for growth, development, and reproduction. Understanding the structure and function of chromosomes sheds light on the inheritance of traits and the transmission of genetic information across generations, elucidating the mechanisms that underpin the marvels of life's diversity and complexity.

MENDELIAN GENETICS

A comprehensive overview of Mendel's laws of inheritance: the law of segregation and the law of independent assortment.

Gregor Mendel, often hailed as the father of modern genetics, formulated two fundamental laws of inheritance through his groundbreaking experiments with pea plants in the mid-19th century. The first law, known as the law of segregation, postulates that during gamete formation, the two alleles (gene variants) for a trait segregate from each other, ensuring that each gamete carries only one allele. Upon fertilization, when gametes unite to form a zygote, the two alleles are once again brought together, restoring the diploid state and providing the genetic basis for the observed traits in the offspring. Mendel's second law, the law of independent assortment, states that genes for different traits assort independently during gamete formation. This principle holds true when genes are located on different chromosomes or when they are located far apart on the same chromosome, thus allowing for various combinations of traits in the offspring. Mendel's laws of inheritance laid the groundwork for understanding the heredity of traits and provided the key principles that later formed the basis of modern genetics, revolutionizing our understanding of how traits are passed from one generation to the next.

Illustration of genetic crosses for monohybrid and dihybrid traits

Punnett squares are powerful tools in genetics that visually depict the possible combinations of alleles resulting from genetic crosses. In a monohybrid cross, involving a single gene with two different alleles, such as "A" and "a," the Punnett square shows the potential genotypes of the offspring. For example, when crossing two individuals with the genotypes "AA" and "aa," all the offspring will have the genotype "Aa," demonstrating the dominant "A" allele's ability to mask the recessive "a" allele. In a dihybrid cross, considering two genes with two alleles each, such as "AaBb" and "AaBb," the Punnett square reveals the combinations of alleles in the offspring. Here, the square is expanded to show four columns and four rows, representing the four possible allele combinations: "AB," "Ab," "aB," and "ab." The dihybrid cross demonstrates how alleles from different genes assort independently, leading to diverse genotypes in the progeny. Punnett squares offer a clear and concise way to understand the expected genetic outcomes of various crosses, aiding researchers and students in predicting and interpreting the inheritance of traits in genetics.

GENETIC VARIATION AND MUTATION

Pedigree analysis as a tool to study the inheritance of genetic traits in families

Pedigree analysis is a powerful and invaluable tool used by geneticists to study the inheritance patterns of genetic traits within families across generations. Represented by a standardized diagram, a pedigree visually depicts the relationships between family members and the presence or absence of specific traits of interest. By examining the patterns of inheritance within a pedigree, geneticists can identify whether a trait is inherited in a dominant, recessive, or X-linked manner. Pedigree analysis provides insights into the genetic basis of various disorders and traits, enabling the identification of carriers, affected individuals, and the likelihood of passing on specific alleles to offspring. This method is particularly useful in the study of inherited diseases, such as cystic fibrosis, Huntington's disease, and hemophilia, as well as for identifying carriers of recessive traits in families with a history of genetic disorders. Through pedigree analysis, genetic counselors and researchers can provide families with crucial information about the risks and inheritance patterns associated with specific genetic traits, ultimately aiding in making informed decisions about healthcare and family planning.

Sources of genetic variation, such as mutations, recombination, and sexual reproduction

Genetic variation is the driving force behind the diversity observed within and among species, providing the raw material for evolution and adaptation. Several sources contribute to the generation of genetic variation. Mutations are one of the primary sources, involving changes in the DNA sequence due to errors during replication, exposure to mutagens, or natural processes. These alterations can result in new alleles, affecting traits and potentially leading to phenotypic differences. Recombination, occurring during meiosis, is another significant source of variation. Homologous chromosomes exchange genetic material, leading to the shuffling of alleles and the creation of unique combinations in the offspring. Sexual reproduction itself introduces variation, as offspring inherit a combination of genes from both parents, leading to diverse genetic profiles. Together, these sources of genetic variation foster the adaptability of organisms to changing environments and drive the course of evolution over time.

Different types of mutations, including point mutations, insertions, deletions, and chromosomal rearrangements

Mutations are alterations in the DNA sequence that can lead to genetic variation and sometimes cause genetic disorders. Point mutations are the most common type, involving a single nucleotide change in the DNA, which can result in substitutions of one base for another (missense or nonsense mutations), or the insertion or deletion of a single nucleotide (frame shift mutations). Insertions and deletions can also involve the addition or removal of multiple nucleotides, disrupting the reading frame of the gene and affecting the protein's function. Chromosomal rearrangements are larger-scale mutations that involve changes in the structure or number of chromosomes. These can include inversions, where a segment of a chromosome is flipped, duplications, where a section of a chromosome is repeated, deletions, where a segment of a chromosome is lost, and translocations, where a piece of one chromosome is transferred to another non-homologous chromosome. These mutations can have varying effects on gene function, and some may lead to genetic diseases or play a role in the development of cancer. Understanding the types and consequences of mutations is crucial for diagnosing genetic disorders, developing therapies, and gaining insights into the complexities of genetics and evolution.

Explain the consequences of mutations on gene function, ranging from neutral variations to deleterious and beneficial mutations.

Mutations can have diverse consequences on gene function, ranging from neutral variations to deleterious and beneficial effects. Neutral mutations have little or no impact on the organism's phenotype or fitness and often occur in non-coding regions of the genome. They can accumulate over time and serve as genetic markers for studying population genetics and evolutionary history. Deleterious mutations, on the other hand, can lead to impaired gene function or the production of non-functional proteins, potentially causing genetic disorders or reducing an individual's fitness. Such mutations are subject to natural selection, with the affected individuals being less likely to survive and reproduce. In contrast, beneficial mutations confer advantages to an organism, enhancing its fitness in specific environments. These mutations can result in new beneficial traits, leading to adaptation and potentially driving the evolution of a population. Overall, the consequences of mutations on gene function are central to understanding the dynamics of genetic diversity, disease inheritance, and the mechanisms that shape the rich tapestry of life's complexity and adaptability.

GENETIC DISORDERS

Categorize genetic disorders based on their inheritance patterns (autosomal dominant, autosomal recessive, X-linked, etc.).

G enetic disorders can be categorized based on their inheritance patterns, which are determined by the distribution of affected genes on chromosomes. Autosomal dominant disorders result from a mutation in one copy of an autosomal gene (non-sex chromosome) and are expressed in individuals with just one mutated allele, usually from an affected parent. Each affected individual has a 50% chance of passing the mutated gene to their offspring. Autosomal recessive disorders, on the other hand, require two copies of the mutated gene, one from each parent, for the disorder to manifest. Carriers, individuals with one normal and one mutated allele, do not show symptoms but can pass on the mutation to their offspring. X-linked disorders arise from mutations in genes located on the X chromosome. Males, who have only one X chromosome, are more commonly affected, as they inherit the mutation from their mothers. Females are usually carriers, with two X chromosomes, and the presence of a normal X chromosome often provides some protection against the disorder. Understanding the inheritance patterns of genetic disorders is crucial for genetic counseling, family planning, and providing appropriate medical care and support to affected individuals and their families.

Examples of well-known genetic disorders like cystic fibrosis, sickle cell anemia, and Huntington's disease, explaining their genetic basis and clinical manifestations.

Cystic fibrosis, sickle cell anemia, and Huntington's disease are well-known genetic disorders that illustrate various inheritance patterns and have distinct genetic bases and clinical manifestations. Cystic fibrosis is an autosomal recessive disorder caused by mutations in the CFTR gene, resulting in the production of defective chloride channels. As a consequence, thick mucus accumulates in the lungs and digestive system, leading to chronic respiratory infections, impaired lung function, and digestive problems. Sickle cell anemia is also an autosomal recessive disorder caused by mutations in the HBB gene, leading to the production of abnormal hemoglobin. The altered hemoglobin causes red blood cells to adopt a sickle-like shape, impairing their ability to carry oxygen, leading to anemia, painful vaso-occlusive crises, and organ damage. Huntington's disease, in contrast, is an autosomal dominant disorder associated with a CAG trinucleotide repeat expansion in the HTT gene. The expanded repeat results in the production of a toxic protein, leading to progressive neurological degeneration, characterized by motor disturbances, cognitive decline, and psychiatric symptoms. These examples demonstrate the diverse genetic basis and clinical manifestations of genetic disorders, highlighting the importance of genetic testing, early diagnosis, and targeted treatments in managing these conditions and providing support to affected individuals and their families.

Importance of genetic counseling and testing for individuals and families affected by genetic disorders

Genetic counseling and testing play a pivotal role in empowering individuals and families affected by genetic disorders with essential information and support Genetic counseling provides a personalized and comprehensive assessment of the genetic risk, inheritance patterns, and potential consequences of genetic disorders. It enables individuals to make informed decisions regarding family planning, medical management, and available treatment options. For families with a history of genetic disorders, genetic testing can help identify carriers and affected individuals, allowing for early diagnosis and intervention, which may improve health outcomes and quality of life. Moreover, genetic counseling fosters open communication about genetic risks and empowers individuals to understand the implications of test results, ensuring they are equipped to make educated choices based on their values and preferences. By facilitating understanding and offering emotional support, genetic counseling and testing provide individuals and families with the tools needed to navigate the complexities of genetic disorders, fostering informed decision-making and promoting a proactive approach to managing their genetic health.

HUMAN GENOME PROJECT AND GENOMICS

History and objectives of the Human Genome Project, emphasizing the completion of the human genome sequence

The Human Genome Project (HGP) was a groundbreaking international scientific endeavor launched in 1990 with the ambitious goal of deciphering the complete sequence of the human genome. It was a collaborative effort involving researchers from around the world, driven by the vision of unraveling the genetic blueprint of humanity. The HGP sought to determine the exact order of the approximately 3 billion base pairs that make up human DNA and identify all the genes present in the genome. This monumental project held great promise for understanding the genetic basis of human health and disease, as well as shedding light on our evolutionary history and the genetic diversity within our species. After more than a decade of dedicated work, the Human Genome Project achieved a historic milestone with the publication of a draft human genome sequence in 2001 and the completion of the high-quality reference human genome sequence in 2003. This achievement marked a turning point in genetics and biomedical research, unleashing an era of genomics that has since revolutionized medicine, personalized healthcare, and our understanding of the intricate genetic underpinnings of life.

Introducing the concept of genomics, which involves the study of entire genomes and their interactions

Genomics is a transformative field of study that focuses on the comprehensive analysis of entire genomes—the complete set of an organism's genetic material. It encompasses the investigation of genes, non-coding DNA, regulatory elements, and their interactions, allowing researchers to understand the molecular and functional aspects of an organism at the genetic level. By examining the entire genomic landscape, genomics provides a holistic perspective on how genes work together to orchestrate complex biological processes, including development, metabolism, and response to the environment. Advancements in high-throughput DNA sequencing technologies have revolutionized genomics, enabling the rapid and cost-effective analysis of vast amounts of genetic data from diverse species. Genomics plays a pivotal role in various scientific disciplines, including medicine, agriculture, ecology, and evolutionary biology, driving discoveries that hold the promise of improving human health, understanding biodiversity, and addressing pressing global challenges.

Exploring the applications of genomics in medicine, including personalized medicine, pharmacogenomics, and disease risk assessment

Genomics has opened up transformative possibilities in medicine, revolutionizing how we diagnose, treat, and prevent diseases. One key application is personalized medicine, where genomic information is used to tailor medical treatments to an individual's unique genetic makeup, maximizing efficacy while minimizing adverse effects. Pharmacogenomics, a subset of personalized medicine, focuses on how an individual's genetic variants influence their response to medications, enabling doctors to prescribe drugs that are most likely to be effective and safe for each patient. Moreover, genomics plays a crucial role in disease risk assessment, identifying genetic markers associated with increased susceptibility to various diseases, such as cancer, cardiovascular disorders, and neurodegenerative conditions. This information empowers individuals to make proactive lifestyle changes and undergo appropriate screenings for early detection, thereby reducing disease burden. Genomics continues to advance our understanding of complex diseases and their genetic basis, paving the way for more targeted therapies, improved patient outcomes, and a transformation in the practice of medicine towards more precise, personalized, and preventive approaches.

MOLECULAR GENETICS

Processes of transcription and translation, highlighting the roles of RNA polymerase, mRNA, tRNA, and ribosomes

Transcription and translation are fundamental processes in molecular biology that work together to convert genetic information into functional proteins. Transcription occurs in the cell nucleus and involves the synthesis of mRNA (messenger RNA) from a DNA template. The enzyme RNA polymerase binds to the promoter region of the gene, initiating the unwinding of the DNA double helix and the synthesis of a complementary mRNA strand using nucleotides. Once the mRNA is transcribed, it undergoes processing, including the addition of a 5' cap and a 3' poly-A tail, and the removal of introns, yielding a mature mRNA that can leave the nucleus and enter the cytoplasm. In the cytoplasm, translation takes place at the ribosomes, where the mRNA is read by the ribosome, and the genetic code is translated into a sequence of amino acids with the help of tRNA (transfer RNA) molecules. Each tRNA carries a specific amino acid and has an anticodon that pairs with the codons on the mRNA. As the ribosome moves along the mRNA, tRNAs bring the corresponding amino acids, and peptide bonds are formed between adjacent amino acids, forming a polypeptide chain. This process continues until a stop codon is reached, and the newly synthesized protein is released. Transcription and translation are intricate and tightly regulated processes that ensure the accurate synthesis of proteins, playing a pivotal role in cell function, development, and the expression of genetic information.

The genetic code, explaining how codons specify amino acids during translation

The genetic code is the set of rules that determines how the sequence of nucleotides in mRNA is translated into the sequence of amino acids in a protein during the process of translation. The genetic code is read in sets of three nucleotides called codons. Each codon specifies a particular amino acid or serves as a start or stop signal for translation. There are 64 possible codons (4 nucleotides arranged in groups of 3), but only 20 amino acids that make up proteins. This means that multiple codons can code for the same amino acid, a property known as degeneracy. The specific correspondence between codons and amino acids is nearly universal among all living organisms, underscoring the shared evolutionary history of life on Earth. For example, the codon AUG codes for the amino acid methionine and serves as the start codon, initiating protein synthesis. As the ribosome reads the mRNA, each codon is matched with the appropriate tRNA carrying the corresponding amino acid. This precise pairing between codons and amino acids is crucial for the accurate and efficient translation of genetic information into functional proteins, which are the building blocks of life and key players in a myriad of cellular processes.

The regulation of gene expression at transcriptional and post-transcriptional levels, including transcription factors, RNA interference, and epigenetic mechanisms

The regulation of gene expression is a sophisticated and tightly controlled process that ensures the precise and coordinated production of proteins in response to various cellular cues. At the transcriptional level, gene expression can be controlled by transcription factors, which are proteins that bind to specific DNA sequences near the gene's promoter region. Transcription factors can activate or repress gene expression by recruiting or blocking the RNA polymerase complex, thereby modulating the transcription process. Additionally, epigenetic mechanisms play a crucial role in gene regulation by altering the chromatin structure. DNA methylation, histone modifications, and chromatin remodeling can influence gene accessibility, either promoting or suppressing transcription. At the post-transcriptional level, RNA interference (RNAi) plays a critical role in regulating gene expression by degrading or inhibiting the translation of specific mRNA molecules. Small RNA molecules, such as microRNAs (miRNAs) and small interfering RNAs (siRNAs), bind to target mRNA molecules, leading to their degradation or translational repression. These intricate regulatory mechanisms ensure that genes are expressed at the right time and in the right cells, allowing cells to adapt to changing environments and carry out their specific functions efficiently. Dysregulation of these processes can lead to various diseases, emphasizing the importance of understanding and studying gene regulation for advances in medicine and biotechnology.

EPIGENETICS

Epigenetics as the study of heritable changes in gene expression that do not involve alterations in the DNA sequence.

E pigenetics is a fascinating and rapidly evolving field of study that delves into the heritable changes in gene expression that occur without alterations in the underlying DNA sequence. Instead of modifying the genetic code, epigenetic mechanisms involve chemical modifications to the DNA or histone proteins, which can influence how genes are turned on or off. These modifications can result in changes to the chromatin structure, affecting the accessibility of genes to the transcriptional machinery. The most common epigenetic marks include DNA methylation, histone acetylation, methylation, and phosphorylation. Epigenetic changes can be induced by various factors, including environmental exposures, diet, and stress. Importantly, these modifications can be passed down from one generation to the next, potentially influencing the health and traits of offspring. Epigenetics plays a pivotal role in various biological processes, including development, differentiation, and disease, and understanding these epigenetic mechanisms opens up new avenues for exploring the complexities of gene regulation and the interplay between genetics and the environment.

Describing the epigenetic mechanisms, such as DNA methylation, histone modifications, and non-coding RNAs

Epigenetic mechanisms are dynamic and intricate processes that regulate gene expression without altering the DNA sequence itself. DNA methylation is one such mechanism, involving the addition of a methyl group to certain cytosine bases in the DNA. Methylation typically represses gene expression by preventing the transcriptional machinery from accessing the gene. Histone modifications, on the other hand, involve chemical alterations to the histone proteins around which DNA is wrapped. These modifications can either promote or inhibit gene expression, depending on the specific modification and its location on the histone tails. Acetylation of histones, for example, often correlates with active gene expression, while methylation can lead to either activation or repression, depending on the context. Another significant group of epigenetic regulators is non-coding RNAs, such as microRNAs (miRNAs) and long non-coding RNAs (lncRNAs). These RNAs can bind to target mRNA molecules, either promoting their degradation or inhibiting their translation, effectively controlling gene expression post-transcriptionally. Together, these epigenetic mechanisms orchestrate the intricate regulation of gene expression, playing essential roles in development, cellular differentiation, and disease pathogenesis. Understanding epigenetic processes has profound implications for unraveling the complexities of gene regulation and its contribution to health and disease.

Role of epigenetics in development, aging, and disease, including cancer and neurodevelopmental disorders

Epigenetics plays a critical role in shaping the trajectory of development, influencing aging processes, and contributing to various diseases, including cancer and neurodevelopmental disorders. During development, epigenetic mechanisms guide the activation and silencing of specific genes at different stages, directing cell differentiation and tissue specialization. As organisms age, epigenetic changes accumulate, leading to altered gene expression patterns and cellular function, which can contribute to age-related diseases and decline. In cancer, aberrant epigenetic modifications can lead to the dysregulation of genes involved in cell growth and proliferation, promoting uncontrolled cell division and tumor formation. Epigenetic factors also underpin the pathogenesis of neurodevelopmental disorders, influencing the formation and function of neural circuits and impacting brain development and function. Understanding the role of epigenetics in these processes provides valuable insights into the underlying molecular mechanisms of development, aging, and disease, offering potential avenues for therapeutic interventions and preventive strategies to mitigate the impact of epigenetic dysregulation on human health.

GENETIC ENGINEERING AND BIOTECHNOLOGY

Principles of genetic engineering, including recombinant DNA technology and gene editing tools like CRISPR/Cas9.

Genetic engineering is a revolutionary discipline that involves the deliberate manipulation of an organism's genetic material to introduce specific traits or modifications. The core principle of genetic engineering relies on recombinant DNA technology, where DNA from different sources is combined to create novel genetic sequences. This is achieved by using restriction enzymes to cut DNA at specific sites, allowing for the insertion of desired genes into a target organism's genome. Recombinant DNA technology has enabled the production of genetically modified organisms (GMOs) with improved agricultural traits, pharmaceutical proteins, and biotechnological applications. Gene editing tools, like CRISPR/Cas9, have further revolutionized genetic engineering. CRISPR/Cas9 allows for precise and efficient modification of specific DNA sequences, enabling the targeted addition, deletion, or correction of genetic information. This technology has vast implications for gene therapy, disease treatment, and fundamental research, offering unprecedented opportunities to explore the function of genes and their role in various biological processes.

However, the ethical considerations and potential risks associated with genetic engineering continue to be subjects of intense debate, emphasizing the need for responsible and informed use of these powerful tools for the betterment of society.

Applications of genetic engineering in medicine (gene therapy), agriculture (GMOs), and industry (biopharmaceuticals)

Genetic engineering has had a profound impact on various sectors, including medicine, agriculture, and industry. In medicine, gene therapy stands as a promising approach to treat genetic disorders by delivering therapeutic genes directly into a patient's cells to correct or replace faulty genes. This technique shows potential in treating inherited diseases like cystic fibrosis and muscular dystrophy, as well as in combating certain types of cancer. In agriculture, the development of genetically modified organisms (GMOs) has revolutionized crop production, leading to plants with enhanced traits such as resistance to pests, diseases, or adverse environmental conditions, ultimately increasing crop yields and food security. Moreover, in the pharmaceutical industry, genetic engineering has facilitated the production of valuable biopharmaceuticals through engineered microorganisms or mammalian cell lines, enabling the large-scale production of complex proteins like insulin, vaccines, and monoclonal antibodies. These applications underscore the far-reaching implications of genetic engineering in improving human health, addressing global food challenges, and advancing industrial processes for a more sustainable and innovative future. However, the responsible and ethical use of genetic engineering remains paramount to address potential risks and ensure the safe and equitable application of these powerful technologies.

Address the ethical considerations surrounding genetic engineering and potential risks and benefits.

Genetic engineering is a revolutionary field with immense potential to address human health, agricultural challenges, and industrial needs. However, it also raises significant ethical concerns that demand careful scrutiny. One key concern lies in the potential unintended consequences of genetic modifications, both in organisms and ecosystems, as well as the potential for unintended spread of engineered genes in the environment. Additionally, the possibility of creating designer babies or enhancing human traits through genetic manipulation raises ethical questions about the limits of human intervention in the natural course of evolution. Furthermore, the use of genetic engineering in agriculture and food production sparks debates about the long-term effects on biodiversity and potential health risks associated with genetically modified organisms (GMOs). Striking a balance between the benefits of genetic engineering and the potential risks requires thoughtful consideration, stringent safety measures, and transparent and inclusive decision-making processes. Robust regulatory frameworks and public engagement are essential to navigate the ethical complexities of genetic engineering, ensuring that its applications are guided by social responsibility, sustainability, and a commitment to the well-being of both current and future generations.

POPULATION GENETICS

Concept of a gene pool and explain how allele frequencies are calculated in populations.

The gene pool refers to the total set of genetic information and allelic variants present in a population of interbreeding individuals. It represents the genetic diversity within the population and serves as the reservoir from which new generations inherit their genetic traits. Allele frequencies, which measure the relative abundance of different alleles in a population, are essential for understanding the genetic makeup and dynamics of a population. To calculate allele frequencies, geneticists count the number of occurrences of each allele for a specific gene locus within a sample of individuals from the population. The frequency of a particular allele is then determined by dividing the number of copies of that allele by the total number of alleles at that gene locus in the population. For instance, if there are 40 copies of allele A and 60 copies of allele B in a sample of 100 alleles, the allele frequency for A would be 40/100 or 0.4 (40%), while the allele frequency for B would be 60/100 or 0.6 (60%). This information provides critical insights into population genetics, evolutionary processes, and the potential for genetic variation within a given population.

Hardy-Weinberg equilibrium as a model for studying allele frequencies in non-evolving populations

The Hardy-Weinberg equilibrium is a fundamental model in population genetics that provides a valuable framework for studying allele frequencies in non-evolving populations. It assumes certain idealized conditions, such as large population size, random mating, no mutation, no migration, and no natural selection. In such a population, the frequencies of alleles at a specific gene locus remain constant over generations. According to the Hardy-Weinberg principle, for a single gene with two alleles (A and a), the frequency of allele A (p) plus the frequency of allele a (q) will always equal 1. Additionally, the genotypic frequencies can be calculated using the equations p^2 for the frequency of homozygous AA individuals, $2pq$ for the frequency of heterozygous Aa individuals, and q^2 for the frequency of homozygous aa individuals. Deviations from the Hardy-Weinberg equilibrium indicate the presence of evolutionary forces at play, such as genetic drift, migration, mutation, or natural selection. By comparing observed allele frequencies in real populations to those expected under the Hardy-Weinberg equilibrium, researchers can identify factors influencing genetic variation and gain insights into the evolutionary dynamics shaping populations over time.

Mechanisms of evolution, including natural selection, genetic drift, gene flow, and mutation, and their impact on genetic diversity

The mechanisms of evolution, namely natural selection, genetic drift, gene flow, and mutation, collectively shape the genetic diversity within populations and drive the gradual change of species over time. Natural selection acts as a powerful force, favoring individuals with advantageous traits that confer higher fitness, leading to the accumulation of beneficial alleles in a population. On the other hand, genetic drift is a random process that can have a significant impact on small populations, leading to fluctuations in allele frequencies by chance alone. Gene flow, the exchange of genetic material between different populations, can introduce new alleles or alter existing ones, promoting genetic diversity and preventing genetic isolation. Mutation, as the ultimate source of genetic variation, introduces new alleles into a population, contributing to the raw material upon which natural selection and other evolutionary processes act. These mechanisms collectively influence the genetic makeup of populations, allowing them to adapt to changing environments, and giving rise to the diversity of life observed on our planet. Understanding the interplay between these evolutionary forces helps elucidate the mechanisms underpinning the rich tapestry of genetic diversity seen in all living organisms.

GENETIC COUNSELING AND ETHICAL CONSIDERATIONS

Role of genetic counselors in helping individuals and families understand and cope with genetic information.

G enetic counselors play a crucial role in helping individuals and families navigate the complexities of genetic information, empowering them to make informed decisions about their health and well-being. With expertise in genetics and counseling, these professionals provide compassionate and personalized support throughout the genetic testing process. Genetic counselors help individuals understand the implications of genetic test results, discuss the risk of developing genetic conditions, and offer guidance on available preventive measures or treatment options. They assist families in interpreting complex genetic information, addressing emotional concerns, and making choices about family planning. Genetic counselors serve as advocates, ensuring that patients' values and preferences are respected, and help facilitate communication between healthcare providers and individuals seeking genetic services. By offering empathetic guidance and evidence-based information, genetic counselors provide invaluable support, helping individuals and families to cope with genetic information and empowering them to make informed decisions about their genetic health.

Ethical issues related to genetic testing, privacy concerns, and the potential for discrimination based on genetic information.

Genetic testing brings significant ethical considerations, particularly regarding privacy and discrimination. As technology advances, access to an individual's genetic information becomes more readily available, raising concerns about data security and confidentiality. The potential for unauthorized access or misuse of genetic data may compromise individuals' privacy, leading to anxiety and reluctance to undergo genetic testing. Moreover, genetic information has the potential to reveal sensitive health conditions or predispositions, making individuals vulnerable to discrimination in various domains, including employment, insurance coverage, and social interactions. Fear of discrimination may deter individuals from seeking genetic testing or disclosing their results, depriving them of valuable medical insights and preventive measures. Striking a balance between the benefits of genetic testing and safeguarding individuals' rights and privacy is paramount, requiring robust policies and regulations to protect genetic information and prevent discriminatory practices, ensuring that genetic testing remains a tool for informed decision-making and improved healthcare without compromising the dignity and well-being of individuals.

Ethical dilemmas of gene editing, germline gene therapy, and human cloning

Gene editing, germline gene therapy, and human cloning raise profound ethical dilemmas that intersect with deeply held moral, social, and philosophical beliefs. Gene editing technologies like CRISPR/Cas9 offer unprecedented opportunities to modify the human genome, potentially curing genetic diseases and improving human health. However, the prospect of heritable changes to the germline raises concerns about the potential for unintended consequences and unknown long-term effects on future generations. Germline gene therapy also raises questions about the line between therapeutic and enhancement interventions, and the possibility of "designer babies" or altering inheritable traits. Additionally, the concept of human cloning evokes ethical considerations regarding the dignity and autonomy of cloned individuals, as well as the potential for exploitation and commodification of human life. These technologies prompt fundamental questions about what it means to be human, the responsible use of scientific advancements, and the need for thoughtful societal dialogue and regulations to ensure that the pursuit of these technologies is guided by ethical principles and respect for human rights and values.

EVOLUTIONARY GENETICS

How genetic variation and natural selection drive the process of evolution

Genetic variation and natural selection are driving forces that fuel the process of evolution. Genetic variation arises from random mutations, recombination during meiosis, and gene flow, leading to differences in individuals' genetic makeup within a population. Natural selection acts upon this variation, favoring traits that provide a survival advantage in specific environments. Individuals with advantageous traits are more likely to survive and reproduce, passing their beneficial genes to the next generation. Over time, the frequency of advantageous alleles increases, while less beneficial or harmful alleles are eliminated. This leads to the adaptation of populations to their changing environments and the evolution of new species. The interplay between genetic variation and natural selection enables organisms to adapt and thrive, driving the diversification and complexity of life on Earth over millions of years.

Genetic evidence for evolution, such as DNA sequencing and comparative genomics

Genetic evidence has provided compelling insights into the evolutionary history of humans, shedding light on our origins and ancient relationships with other species. DNA sequencing and comparative genomics have played pivotal roles in this endeavor. By comparing the genomes of modern humans with those of our closest relatives, such as Neanderthals and Denisovans, scientists have identified shared genetic material, indicating interbreeding events and gene flow between ancient human populations. Moreover, studying the genetic diversity within human populations has provided evidence for migration patterns and the colonization of different regions of the world. Additionally, the study of ancient DNA from fossilized remains has allowed researchers to reconstruct the genetic landscape of our ancestors and unravel the genetic adaptations that accompanied human evolution. The growing field of genetic evidence has provided remarkable insights into the complex and dynamic evolutionary journey of humanity, further enriching our understanding of the interconnectedness of all living beings and the fascinating story of our species' past.

Impact of evolutionary genetics on our understanding of human history and biodiversity

Evolutionary genetics has revolutionized our understanding of human history and biodiversity, providing crucial insights into the complex interplay of genes, natural selection, and migration that shaped the course of human evolution. By analyzing genetic data from diverse populations, researchers have unveiled migration patterns, colonization routes, and demographic shifts that have played pivotal roles in human history. The study of ancient DNA has revealed our connections with long-extinct hominin species and illuminated the interactions and interbreeding events that occurred throughout prehistoric times. Furthermore, evolutionary genetics has enriched our understanding of biodiversity by elucidating the genetic relationships between different species and populations, helping us grasp the underlying mechanisms that drive adaptation and speciation. This growing field continues to uncover the deep-rooted connections between humans and other organisms, elucidating the intricate tapestry of life's diversity and providing valuable insights into our shared ancestry and the dynamic processes that have shaped the living world we inhabit today.

CANCER GENETICS

Genetic basis of cancer, including oncogenes, tumor suppressor genes, and chromosomal abnormalities

The genetic basis of cancer is characterized by alterations in key genes that regulate cell growth and division. Oncogenes are genes that, when mutated or overexpressed, promote uncontrolled cell proliferation, contributing to tumor formation. These mutations can lead to the activation of growth-promoting signaling pathways, allowing cells to bypass normal growth control mechanisms. Conversely, tumor suppressor genes act as guardians of the genome, regulating cell cycle progression and DNA repair. Inactivating mutations or deletions of tumor suppressor genes can lead to the loss of their normal function, allowing damaged cells to escape surveillance and proliferate uncontrollably. Additionally, chromosomal abnormalities, such as translocations, deletions, and amplifications, can cause gene dysregulation and contribute to the development of cancer. These genetic changes collectively drive the development and progression of cancer, highlighting the intricate interplay between genetic alterations and cellular processes that underlie this complex disease. Understanding the genetic basis of cancer has significant implications for cancer diagnosis, treatment, and the development of targeted therapies aimed at disrupting the molecular mechanisms driving cancer growth and survival.

Difference between somatic mutations and germline mutations in cancer development

The key distinction between somatic mutations and germline mutations in cancer development lies in their origins and impact on heredity. Somatic mutations occur in non-reproductive cells during an individual's lifetime and are not passed on to offspring. These mutations are responsible for the development of most cancers and are often caused by various environmental factors, such as exposure to carcinogens or errors in DNA replication and repair. Somatic mutations are specific to the affected individual and are present only in the tumor cells, leading to the growth and progression of cancer. In contrast, germline mutations occur in the egg or sperm cells and are passed on to subsequent generations. These mutations can be inherited from a parent and are present in all cells of the offspring, potentially increasing the risk of developing certain types of cancer at an early age. While both somatic and germline mutations contribute to cancer development, understanding their differences is crucial for accurate diagnosis, genetic counseling, and targeted treatment strategies in cancer patients and their families.

Hereditary cancer syndromes and the role of genetic testing in cancer risk assessment

Hereditary cancer syndromes are characterized by an increased risk of developing specific types of cancer due to inherited genetic mutations. These mutations are present in the germline and can be passed down from one generation to the next. Individuals with hereditary cancer syndromes have a higher likelihood of developing cancer at an earlier age and may have multiple family members affected by the same or related cancers. Genetic testing plays a pivotal role in cancer risk assessment for individuals and families with a suspected hereditary predisposition to cancer. Testing can identify specific gene mutations associated with hereditary cancer syndromes, such as BRCA1 and BRCA2 in hereditary breast and ovarian cancer or Lynch syndrome in hereditary colorectal and endometrial cancer. The results of genetic testing inform personalized cancer risk assessment, enabling targeted surveillance and prevention strategies, such as increased screening or risk-reducing surgeries. Moreover, genetic testing helps identify at-risk family members, allowing them to undergo appropriate testing and take proactive measures to manage their cancer risk. Early detection and intervention based on genetic testing results have the potential to save lives and significantly impact cancer prevention and management in individuals and families with hereditary cancer syndromes.

MITOCHONDRIAL GENETICS

Unique features of mitochondrial DNA, such as its maternal inheritance and lack of recombination.

Mitochondrial DNA (mtDNA) possesses unique features that set it apart from nuclear DNA. One of its most distinctive attributes is maternal inheritance, where mtDNA is passed down exclusively from the mother to her offspring. This pattern of inheritance allows researchers to trace maternal lineages across generations and study human population history and migration patterns. Another striking characteristic of mtDNA is its lack of recombination. Unlike nuclear DNA, which undergoes recombination during meiosis, mtDNA is inherited as a single, intact unit. This absence of recombination preserves the genetic information accumulated through generations, providing a valuable source of information for evolutionary studies. The combination of maternal inheritance and the absence of recombination in mtDNA offers insights into human evolution, population genetics, and maternal lineages, making it a valuable tool in various fields of research.

Mitochondrial diseases and their association with mutations in mitochondrial DNA

Mitochondrial diseases are a group of inherited disorders characterized by dysfunctional mitochondria, the cellular powerhouses responsible for producing energy in the form of adenosine triphosphate (ATP). These disorders are associated with mutations in mitochondrial DNA (mtDNA), which encode essential components of the mitochondrial respiratory chain responsible for ATP production. Since mitochondria are inherited exclusively from the mother, mitochondrial diseases often exhibit a maternal inheritance pattern. The severity and symptoms of mitochondrial diseases vary widely, affecting multiple organ systems, including the brain, muscles, heart, and liver. Common clinical features include muscle weakness, neurological problems, developmental delays, and metabolic abnormalities. The large proportion of energy-dependent processes in the body makes mitochondrial dysfunction particularly impactful, leading to a wide range of clinical presentations and challenges in diagnosis and treatment. Advances in genetic testing and our understanding of mitochondrial DNA mutations have improved diagnostic accuracy, enabling tailored management strategies and family counseling for those affected by these challenging and often debilitating disorders.

Role of mitochondria in energy production and their involvement in cellular signaling

Mitochondria play a dual role in cellular function, serving as the primary sites of energy production and as key players in cellular signaling pathways. As the powerhouses of the cell, mitochondria carry out aerobic respiration, a process that converts nutrients, such as glucose and fatty acids, into ATP, the molecule that stores and transfers energy within cells. This highly efficient process occurs within the inner mitochondrial membrane, where electron transport and ATP synthesis take place. Additionally, mitochondria are involved in cellular signaling through the release of signaling molecules, calcium ions, and reactive oxygen species (ROS). Mitochondrial signaling influences various cellular processes, including cell survival, apoptosis (programmed cell death), and stress responses. The dynamic interplay between energy production and signaling functions underscores the crucial role of mitochondria in maintaining cellular homeostasis and regulating cell fate, making them indispensable organelles for the proper functioning of living organisms.

COMPARATIVE GENOMICS

Importance of comparative genomics in understanding evolutionary relationships between species

Comparative genomics has emerged as a powerful tool for understanding evolutionary relationships between species, providing valuable insights into the shared ancestry and divergence of organisms across the tree of life. By comparing the genetic sequences and genomic structures of different species, researchers can identify conserved genes and regions that have remained largely unchanged throughout evolution, as well as lineage-specific genetic changes that have contributed to species diversification. Comparative genomics has revealed the patterns of gene gain, loss, and duplication that underlie the evolution of novel traits and adaptations in different lineages. It has also illuminated the genetic basis of phenotypic diversity and functional innovations, shedding light on the processes that have shaped the remarkable biodiversity seen on Earth today. Moreover, comparative genomics allows for the identification of genomic regions under positive selection or subject to functional constraints, offering insights into the genetic underpinnings of species-specific traits and ecological adaptations. The exploration of shared and unique genomic features among different species helps researchers understand the genetic mechanisms driving evolutionary changes, providing a comprehensive and in-depth understanding of the history of life on our planet.

Use of model organisms in genetic research, such as mice, fruit flies, and yeast

Model organisms, such as mice, fruit flies, and yeast, have played a pivotal role in advancing genetic research and our understanding of fundamental biological processes. These organisms offer numerous advantages, including short lifespans, rapid reproduction rates, and ease of manipulation in the laboratory. By selectively breeding and studying genetic variants, researchers can unravel the genetic basis of specific traits and diseases. Model organisms have provided invaluable insights into the molecular mechanisms underlying development, aging, and disease. For example, mice have been instrumental in studying mammalian genetics and human diseases due to their genetic similarity to humans. Fruit flies have served as excellent models for understanding genetic inheritance and developmental processes. Yeast, a single-celled organism, has been essential for elucidating basic cellular processes, including DNA replication and repair. The findings obtained from these model organisms have often been extrapolated to other organisms, including humans, contributing to significant advancements in biomedical research and offering potential targets for therapeutic interventions and genetic therapies.

How comparative genomics can reveal conserved genetic elements and identify candidate genes for human diseases

Comparative genomics has proven to be a valuable approach in revealing conserved genetic elements and identifying candidate genes for human diseases. By comparing the genomes of humans with those of other organisms, such as model organisms or closely related species, researchers can identify regions of the genome that have been preserved throughout evolution, suggesting their functional significance. Conserved genetic elements often represent crucial regulatory regions or functional elements that are essential for normal cellular processes. Additionally, comparative genomics can highlight genes that have been conserved across species, indicating their fundamental importance in biological functions. When researchers identify genes associated with a particular disease in model organisms or other species, these genes become strong candidates for similar disease mechanisms in humans. Furthermore, studying the genetic underpinnings of diseases across species provides insights into the shared genetic pathways and disease mechanisms, leading to the discovery of potential therapeutic targets and the development of treatments for human diseases. The power of comparative genomics lies in its ability to leverage evolutionary relationships to uncover important genetic elements and to bridge the gap between model organisms and human diseases, ultimately enhancing our understanding of human biology and disease.

PHARMACOGENETICS

Genetic basis of drug response and the concept of individual variability in drug metabolism and efficacy

The genetic basis of drug response lies in the variations within individual genomes that influence how drugs are metabolized, interact with target molecules, and elicit therapeutic effects. Genetic variations in drug-metabolizing enzymes, such as cytochrome P450 enzymes, can lead to differences in drug metabolism rates, affecting drug efficacy and toxicity. Similarly, genetic variants in drug target proteins can alter the drug's binding affinity, influencing the response to medication. The concept of individual variability in drug metabolism and efficacy highlights that each person's genetic makeup influences how they respond to drugs, making personalized medicine a promising approach to optimize treatment outcomes. Pharmacogenomics, the study of how genes affect drug response, helps identify genetic markers associated with drug efficacy and safety, enabling healthcare professionals to tailor drug therapies based on a patient's genetic profile. Embracing the genetic basis of drug response allows for more precise and effective treatments, minimizing adverse reactions and enhancing therapeutic outcomes for individual patients.

Use of pharmacogenomic testing to personalize drug treatments and avoid adverse drug reactions

Pharmacogenomic testing has revolutionized personalized medicine by tailoring drug treatments to an individual's genetic makeup, minimizing adverse drug reactions, and optimizing therapeutic outcomes. By analyzing specific genetic markers that influence drug metabolism, efficacy, and safety, pharmacogenomic testing helps identify patients who may have a higher risk of adverse reactions or reduced response to certain medications. Armed with this information, healthcare providers can make more informed decisions about drug selection, dosing, and treatment duration, ensuring that patients receive the most effective and well-tolerated medications. Pharmacogenomic testing is particularly beneficial for drugs with a narrow therapeutic window or significant inter-individual variability in drug metabolism. This approach empowers clinicians to avoid trial and error prescribing and helps prevent potentially harmful side effects, leading to safer and more efficient drug treatments tailored to the unique genetic profile of each patient.

Challenges and potential benefits of integrating pharmacogenetics into clinical practice

Integrating pharmacogenetics into clinical practice presents both challenges and potential benefits. One of the main challenges lies in the complexity of genetic testing and the need for reliable and standardized testing platforms. Implementing pharmacogenetics testing on a large scale requires robust infrastructure, specialized training for healthcare professionals, and efficient methods for result interpretation and incorporation into clinical decision-making. Additionally, the availability of evidence-based guidelines and clear evidence of clinical utility for specific drugs and conditions are essential for the successful integration of pharmacogenetics into routine care. However, the potential benefits are significant. Personalizing drug treatments based on a patient's genetic profile can lead to improved treatment outcomes, increased drug efficacy, and reduced adverse drug reactions. This approach holds promise for enhancing patient safety, reducing healthcare costs by avoiding ineffective or harmful treatments, and improving patient satisfaction and adherence to treatment plans. As research and technology in pharmacogenetics advance, the integration of this valuable information into clinical practice has the potential to transform healthcare, enabling more precise and personalized medicine for the benefit of patients worldwide.

GENETICS AND AGRICULTURE

Principles of selective breeding in agriculture and animal husbandry

Selective breeding, a fundamental practice in agriculture and animal husbandry, involves choosing specific plants or animals with desirable traits to mate and produce offspring with those traits. The process is guided by the principles of heritability, variation, and selection. Heritability refers to the proportion of a trait that is passed on from parent to offspring genetically, ensuring that desirable traits are transmitted to subsequent generations. Variation within a population allows breeders to identify individuals with the desired characteristics, such as increased yield, disease resistance, or specific physical attributes. Through controlled mating, breeders select the best-performing individuals as parents for the next generation, gradually enhancing the prevalence of the desired traits within the population. Over time, this deliberate process of selection and breeding results in improved crop varieties and livestock breeds that are better suited to specific agricultural conditions and human needs, contributing to increased agricultural productivity and sustainability

Development and adoption of genetically modified organisms (GMOs) in agriculture

The development and adoption of genetically modified organisms (GMOs) in agriculture have sparked significant advancements and controversies. GMOs are organisms whose genetic material has been altered using biotechnology techniques to introduce desirable traits or enhance their resistance to pests, diseases, or environmental stressors. In the late 20th century, the advent of genetic engineering allowed researchers to transfer genes between different species, opening up new possibilities for agricultural improvement. GMOs have been widely adopted for crops like soybeans, corn, cotton, and canola, offering increased yields and reduced reliance on pesticides. However, their use has also stirred debates concerning environmental and health impacts, potential crossbreeding with wild populations, and corporate control over seed supply. Regulatory frameworks and labeling requirements vary worldwide, reflecting the ongoing dialogue between the potential benefits and risks associated with GMOs. As technologies evolve, the future of GMOs in agriculture will depend on striking a balance between scientific advancements, ethical considerations, and addressing concerns related to human health, biodiversity, and sustainable agriculture practices.

Controversies and ethical considerations surrounding GMOs and their potential environmental impacts

The use of genetically modified organisms (GMOs) in agriculture has been a subject of intense controversy and ethical deliberation, particularly concerning their potential environmental impacts. Critics express concerns over the unintended consequences of GMOs, such as the potential for crossbreeding with wild plant species, leading to unintended spread of engineered genes and disrupting natural ecosystems. Additionally, there are worries that GMOs may lead to the development of resistant pests and weeds, necessitating increased pesticide use and impacting biodiversity. Ethical considerations encompass issues of corporate control over seed supply, food security in developing countries, and the potential economic dependency on biotechnology companies. The lack of long-term studies on the ecological impacts of GMOs further fuels the debate. Balancing the potential benefits of increased crop yields and food production with environmental sustainability, biodiversity conservation, and the precautionary principle remains a significant challenge, highlighting the need for robust regulations, scientific research, and transparent public engagement to address the complex ethical dimensions surrounding GMOs in agriculture.

GENETICS AND HUMAN EVOLUTION

Genetic evidence for human evolution, including DNA sequencing of extinct hominins

Genetic evidence has provided compelling insights into human evolution, illuminating our ancient connections with extinct hominin species. DNA sequencing of ancient hominin remains, such as Neanderthals and Denisovans, has allowed researchers to reconstruct their genomes and compare them with modern human genomes. These comparisons have revealed interbreeding events and gene flow between ancient hominin populations and early modern humans, highlighting the dynamic interactions and complex evolutionary history of our species. Additionally, studying genetic variation within modern human populations has shed light on migration patterns, population movements, and the genetic diversity of different human groups across the globe. By integrating genetic evidence with fossil and archaeological data, researchers have pieced together a comprehensive narrative of human evolution, deepening our understanding of the shared ancestry and rich evolutionary heritage of Homo sapiens and our extinct hominin relatives. This genetic evidence serves as a powerful tool to unravel the mysteries of our evolutionary past and explore the factors that shaped the diversity and adaptability of our species.

Impact of migration, gene flow, and natural selection on human genetic diversity

Migration, gene flow, and natural selection have had a profound impact on human genetic diversity, shaping the complex mosaic of genetic variation observed among different populations. Migration events, whether driven by climate changes, exploration, or social factors, have facilitated the spread of genetic traits across continents, leading to the mixing and intermingling of previously isolated populations. Gene flow, the exchange of genetic material between populations through interbreeding, continues to influence human genetic diversity by introducing new genetic variants and promoting the sharing of advantageous traits. Moreover, natural selection, the differential survival and reproduction of individuals with certain genetic characteristics, has acted on human populations throughout history, favoring adaptations to local environments, such as disease resistance or tolerance to specific diets. These evolutionary forces have shaped human genetic diversity, reflecting the interplay of historical events, environmental pressures, and the unique cultural and demographic histories of human societies. Studying the patterns of genetic diversity provides valuable insights into human evolution, migration patterns, and the fascinating story of our species' journey across the globe.

Debates and controversies related to human evolutionary history, such as the "Out of Africa" hypothesis

Debates and controversies surrounding human evolutionary history are rooted in the quest to understand our origins and the processes that shaped our species. One prominent debate revolves around the "Out of Africa" hypothesis, which proposes that modern humans originated in Africa and subsequently dispersed to other parts of the world, replacing earlier hominin populations. This hypothesis is supported by genetic evidence showing that the genetic diversity of modern human populations is highest in Africa, indicating an African origin for our species. However, alternative theories, such as the "Multiregional" model, suggest that modern humans evolved simultaneously in different regions and interbred with archaic hominins, leading to their genetic contributions in present-day populations. While the genetic evidence overwhelmingly supports an African origin for Homo sapiens, the debate reflects complex interactions between migration, gene flow, and genetic admixture, which can make untangling the precise details of our evolutionary history challenging. Resolving these debates requires interdisciplinary research, incorporating data from genetics, fossils, archaeology, and other disciplines, to provide a comprehensive understanding of the rich tapestry of human evolutionary history.

CURRENT TRENDS AND FUTURE DIRECTIONS IN GENETICS

Recent breakthroughs in genetics research, such as advances in gene editing, gene therapy, and synthetic biology

In recent years, genetics research has witnessed remarkable breakthroughs, revolutionizing our understanding and application of genetic technologies. Advances in gene editing, particularly with the development of CRISPR/Cas9, have empowered scientists to precisely modify DNA sequences in various organisms, offering unprecedented potential for treating genetic diseases, engineering agricultural crops, and exploring fundamental biological processes. In the realm of gene therapy, researchers have achieved significant successes in treating previously incurable genetic disorders, with FDA-approved treatments for conditions like spinal muscular atrophy and inherited retinal diseases. Additionally, progress in synthetic biology has enabled the design and construction of novel biological systems, paving the way for bio-based solutions in areas like medicine, energy, and environmental remediation. These breakthroughs hold immense promise for tackling previously intractable challenges, shaping the future of genetics research, and opening new frontiers in personalized medicine, biotechnology, and our understanding of life itself.

Potential applications of gene editing technologies, including CRISPR-based treatments for genetic disorders and infectious diseases

Gene editing technologies, with CRISPR/Cas9 at the forefront, hold immense potential for a wide range of applications, particularly in the fields of genetic medicine and infectious disease treatment. In genetic disorders, CRISPR-based treatments offer the prospect of correcting or modifying disease-causing genetic mutations directly at the DNA level. Researchers have made significant strides in preclinical studies, demonstrating the feasibility of using CRISPR to treat genetic conditions like sickle cell anemia, cystic fibrosis, and Duchenne muscular dystrophy. Additionally, gene editing provides a promising avenue for tackling infectious diseases. By precisely targeting viral genomes, CRISPR-based therapies may be used to disable or eliminate pathogenic viruses, such as HIV, hepatitis B, and influenza. Moreover, the development of CRISPR-based antiviral strategies may enable the prevention or treatment of emerging infectious diseases that pose global health threats. As gene editing technologies continue to advance, their potential applications in genetic and infectious disease treatments hold the promise of transforming medical practices and improving the lives of countless individuals worldwide. However, their implementation must be approached with careful consideration of ethical, regulatory, and safety considerations to ensure responsible and equitable use in clinical settings.

Ethical and societal considerations associated with emerging genetic technologies

Emerging genetic technologies raise profound ethical and societal considerations that must be carefully addressed to ensure their responsible and equitable use. As gene editing techniques, such as CRISPR/Cas9, advance, questions arise about the ethical implications of manipulating the human germline, potentially affecting future generations. Striking a balance between the pursuit of medical advancements and the precautionary principle regarding heritable genetic changes becomes essential. Additionally, concerns over genetic discrimination, privacy, and consent in genetic testing and data-sharing require robust regulations to safeguard individuals' rights and autonomy. The availability and affordability of genetic technologies also raise equity and access issues, with potential disparities in healthcare access and unequal distribution of benefits. The broader societal implications include discussions on the definition of disability, concepts of genetic "enhancement," and cultural attitudes toward genetic modifications. Engaging stakeholders, including scientists, policymakers, ethicists, and the public, in transparent discussions is vital for establishing comprehensive ethical guidelines and governance frameworks that navigate the complexities of emerging genetic technologies responsibly, promoting their potential benefits while safeguarding human rights, dignity, and societal values.

THE PHILOSOPHY OF GENETIC STUDIES

Ethical, epistemological, and metaphysical aspects of genetics

Genetics, the study of our hereditary information encoded within DNA, has not only unlocked the secrets of our biological makeup but has also raised profound questions that extend far beyond the laboratory. It is within the intersection of genetics and philosophy that we embark on a journey to explore the ethical, metaphysical, and epistemological dimensions of genetic study. As we venture into the philosophy of genetic research, we will grapple with complex ethical considerations surrounding the manipulation of our genetic code and the implications for future generations. We will contemplate the very essence of identity, exploring the age-old debate of nature versus nurture, and ponder the profound philosophical questions posed by the deterministic aspects of genetic discoveries. In our quest for understanding, we will also examine the epistemology of genetics, addressing the nature of genetic knowledge, its limitations, and the intricate interplay between genetics and philosophy. This exploration will reveal how genetics and philosophy inform one another, creating a rich tapestry of insight that extends beyond the realm of science and touches upon the very essence of what it means to be human. The study of genetics and philosophy, though seemingly distinct fields of inquiry, share a unique intersection that has the power to reshape our understanding of existence. At the core of this convergence lies the ethical imperative to navigate the moral landscape of genetic research. Genetic engineering, gene editing, and the ability to manipulate our very genetic code raise profound ethical dilemmas. Philosophical inquiry guides us in grappling with questions of right and wrong, consent, and the consequences of playing with the very essence of life itself. It challenges us to reflect on our responsibilities as stewards of this knowledge and the potential implications for future generations. Furthermore, the interplay between genetics and philosophy delves into the philosophical questions surrounding human identity. The age-old debate of nature versus nurture is reinvigorated as genetics increasingly unveils the role of our genes in shaping who we are. This debate poses fundamental questions about the essence of human agency and free will in a world where our genetic predispositions are better understood. It beckons us to contemplate the intersection of genetic determinism and personal autonomy, prompting philosophical introspection into the heart of human identity and the enduring conundrum of what it means to be human in an age of genetic revelation. The significance of ethical and philosophical reflection in the realm of genetic study cannot be overstated. It serves as the moral compass guiding our exploration of the genetic code, a roadmap that charts our path through the intricate ethical dilemmas that accompany unprecedented scientific discoveries. In a world where we possess the power to edit, manipulate, and intervene in the very blueprint of life, ethical considerations become paramount. Philosophical inquiry helps us navigate these ethical waters, urging us to contemplate the potential consequences of our actions and to discern between what is scientifically possible and what is ethically responsible. Furthermore, the philosophical dimension in genetic study extends beyond the immediate ethical concerns. It prompts us to ponder the enduring questions of human existence and identity. Through the lens of philosophy, we seek to unravel the intricate interplay between genetics and human agency. This intellectual endeavor allows us to engage with the ethical and metaphysical implications of genetic determinism and to grapple with profound questions about free will, individual autonomy, and the essence of human identity. In doing so, ethical and philosophical reflection in genetic study illuminates not only the path we tread in the realm of science but also the deep philosophical inquiries that are essential in shaping our understanding of life itself.

Ethical Dilemmas in Genetic Engineering and Editing

Genetic engineering and editing have ushered in a new era of scientific possibility, one where we possess the tools to modify the fundamental building blocks of life. Yet, with this remarkable power comes a profound ethical dilemma. At the heart of this dilemma lies the question of where we should draw the line between what we can do and what we should do. The very act of altering the genetic code raises concerns about unintended consequences, unforeseen ripple effects, and the irreversible nature of genetic changes. Ethical reflection prompts us to consider the potential harms and benefits, the risks and rewards, and the ethical responsibilities that accompany the power to manipulate our own genetic heritage. As we delve deeper into the realm of genetic engineering and editing, we confront the delicate issue of consent. The ability to make genetic alterations raises questions about informed consent and the autonomy of individuals, particularly when it comes to editing the genes of future generations. The ethical quandary surrounding germline editing, in which changes are hereditary, forces us to reflect on the implications for the rights of those who cannot provide informed consent, the potential for creating "designer babies," and the role of society in regulating these profound genetic choices. Moreover, the ethical dilemmas surrounding genetic engineering extend to considerations of equity and access. Who gets to benefit from genetic enhancements, and who is left behind? The potential for exacerbating societal disparities and creating a genetic divide between the privileged and the marginalized raises profound ethical concerns. Ethical reflection encourages us to examine the implications of these disparities and to contemplate how to ensure that genetic advancements are used for the betterment of all, rather than perpetuating inequality. Lastly, the ethical dimension of genetic engineering and editing challenges us to consider the broader ecological and environmental consequences. Modifying the genetic code of organisms may lead to unintended ecological disruptions, affecting ecosystems and biodiversity. Ethical dilemmas arise when we weigh the potential benefits of genetic interventions against the ecological risks and unknown long-term effects. This compels us to think deeply about our ethical responsibilities not only to humanity but to the delicate balance of life on our planet. In the face of these profound ethical dilemmas, the intersection of genetics and philosophy calls upon us to engage in a thoughtful, informed, and conscientious exploration of the ethical complexities surrounding genetic engineering and editing. This introspection becomes essential as we navigate the frontiers of science and grapple with the profound implications of our newfound genetic capabilities.

Genetic Privacy and Consent

Genetic privacy and consent have emerged as critical ethical and legal considerations in the age of genetic exploration. The collection and analysis of individuals' genetic data, while promising in terms of personalized medicine and scientific advancement, give rise to profound questions regarding privacy, autonomy, and informed consent. At the core of genetic privacy is the concern that our most intimate biological information, encoded in our DNA, can be shared, stored, and potentially exploited without our knowledge or consent. The potential misuse of genetic information, whether by insurers, employers, or other entities, poses a genuine threat to our individual autonomy. Ethical considerations guide us to safeguard genetic privacy, ensuring that individuals have control over who accesses their genetic data and for what purpose.

Informed consent, a cornerstone of ethical research and medical practice, is equally paramount in the genetic realm.

Individuals must be provided with comprehensive information about the purposes, risks, and potential benefits of genetic testing or research. The right to make informed decisions about participating in genetic studies, sharing one's genetic data, or undergoing genetic testing is not only an ethical principle but a legal requirement in many jurisdictions. Respect for autonomy and self-determination compels researchers and healthcare providers to engage in transparent and comprehensive consent processes, thereby empowering individuals to make decisions that align with their values and interests. The intricate landscape of genetic privacy and consent becomes even more intricate when we consider the context of family and group genetic data. The sharing of genetic information with family members, both known and unknown, can have cascading effects. Ethical dilemmas may arise concerning the disclosure of unexpected or potentially distressing genetic findings to relatives. As genetic testing becomes more common, there is a growing need to address the ethical and legal implications of family genetic data, especially in situations involving hereditary diseases or conditions. Furthermore, the international dimension of genetic privacy and consent cannot be overlooked. In an era of global collaboration and information exchange, genetic data may cross national borders and legal jurisdictions, raising challenging questions about the harmonization of ethical standards, privacy regulations, and data security on a global scale. Striking a balance between the pursuit of scientific knowledge and the preservation of individual and collective rights is an ongoing challenge that necessitates international ethical cooperation. In sum, genetic privacy and consent are foundational to responsible genetic research and medical practice. As genetic technologies continue to advance, it becomes imperative to uphold the principles of privacy and informed consent, ensuring that individuals and their genetic information are treated with the utmost respect, and that the promise of genetics is realized in an ethical and equitable manner.

Nature vs. Nurture: The Role of Genetics in Shaping Identity

The age-old debate of nature versus nurture has found new resonance in the field of genetics, as our understanding of the role of genes in shaping identity deepens. This ongoing philosophical inquiry delves into the question of how much of who we are, as individuals and as a species, can be attributed to our genetic inheritance, and how much is the result of our environmental influences and life experiences. In contemplating the role of genetics in identity, we confront the intriguing notion that our genes provide a genetic roadmap, encoding the potential for a vast array of physical, cognitive, and emotional traits. From our predispositions to certain diseases to the color of our eyes and the structure of our brains, genes lay the foundation for our biological identity. But as philosophers and geneticists alike ponder, the mere presence of genetic information does not predetermine who we become. It is the intricate interplay between genes and the environment that shapes the unique individual identities we each embody. The exploration of genetic determinism, the idea that our genes rigidly dictate our identity, often leads to complex ethical and philosophical considerations. While our genes may play a crucial role in shaping our potential, they do not write a predetermined script for our lives. Rather, the unfolding of our identity is a dynamic narrative, influenced by an ever-changing interplay between genetic heritage and environmental experiences. This philosophical perspective challenges deterministic notions and underscores the significance of personal autonomy in shaping one's identity. Moreover, the role of genetics in identity extends beyond the realm of individual traits to encompass broader questions of human nature and culture. The genetic study of human populations has enriched our understanding of our shared genetic heritage, revealing that we are more alike than different as a species. In this context, genetic identity transcends the individual, offering insights into the collective identity of humanity. As the debate between nature and nurture persists, it reminds us that our understanding of identity is inextricably linked to the mysteries of our genetic inheritance and the complex tapestry of the human experience. In conclusion, the question of nature versus nurture continues to captivate the philosophical imagination as genetics sheds light on the intricate role of genes in shaping our identities. The ongoing dialogue reminds us that our genetic heritage is a rich and complex narrative, a starting point rather than an endpoint in the journey of self-discovery. It prompts us to reflect on the balance between our shared human genetic nature and the unique individual identities we create through the experiences and choices that define our lives.

Determinism vs. Free Will: The Implications of Genetic Discoveries

Genetic discoveries have breathed new life into age-old philosophical debates, particularly the enduring clash between determinism and free will. As we delve into the intricate landscape of genetics, we confront profound implications regarding the extent to which our genes shape our lives and the room left for individual agency and free will. The deterministic aspects of genetic discoveries challenge the traditional notion of free will, as they suggest that our genetic makeup predisposes us to certain traits and behaviors. The mapping of genes associated with mental health conditions, personality traits, and even propensities for addiction raises questions about the degree to which we can consciously control our actions. This philosophical dilemma prompts us to explore the extent to which our decisions and choices are influenced by genetic predispositions, and whether we possess true autonomy over our lives. On the other hand, genetic discoveries do not necessarily lead to a deterministic worldview. Instead, they offer a nuanced perspective that highlights the interaction between genetics and the environment. While our genes may influence our tendencies, it is the interplay between genes and the world around us that ultimately shapes our behaviors and choices. This perspective underscores the complexity of human identity and acknowledges the role of free will in determining the path our lives take. The philosophical debate between determinism and free will gains particular urgency when considering the implications of genetic discoveries in legal and moral contexts. It prompts us to reflect on issues such as criminal responsibility, where genetic predispositions may be cited as factors in legal defenses. Additionally, it challenges the ethical dimensions of genetic interventions, such as gene editing, and the responsibility of individuals and society in wielding the power to modify the genetic code. In essence, the implications of genetic discoveries on the determinism versus free will debate invite us to engage in profound philosophical introspection. They call upon us to examine the balance between the potential genetic influences on our lives and the enduring capacity for human agency. This ongoing dialogue reminds us that genetics is not destiny but a complex thread in the tapestry of human identity, where the interplay of genes and free will is an enduring philosophical mystery.

The Limits of Genetic Determinism

While genetic research offers remarkable insights into the role of genes in human health and behavior, it also draws attention to the inherent limitations of genetic determinism. The concept of genetic determinism suggests that our genes rigidly dictate our traits and behaviors, leaving little room for environmental influence or individual autonomy. However, as we delve deeper into the intricacies of genetics, it becomes increasingly apparent that this deterministic perspective has its bounds. One limitation of genetic determinism lies in the complexity of gene-gene and gene-environment interactions. Genes do not operate in isolation but rather in a network of intricate interactions. They respond to environmental cues, and their effects are often modulated by the presence of other genes. The interplay between multiple genetic factors and environmental conditions results in a dynamic, nonlinear system that defies simple determinism. This complexity underscores the challenge of attributing a single genetic cause to complex traits and behaviors. Moreover, genetic determinism often underestimates the influence of environmental factors in shaping human characteristics. Our genes may provide a genetic predisposition, but environmental exposures and experiences play a significant role in the expression of these traits. This leads to the recognition that while genes set the stage, it is the environment that directs the play. The limits of genetic determinism become evident when we acknowledge the profound impact of external factors, such as upbringing, culture, and life experiences, in molding our identity and behavior. The implications of the limits of genetic determinism are multifaceted. They challenge deterministic notions that our genes are our destiny and emphasize the importance of free will, individual agency, and the capacity for change. In the realm of personalized medicine, recognizing these limits calls for a more holistic understanding of health, considering both genetic and environmental factors. Additionally, it highlights the ethical responsibility to avoid genetic reductionism, which oversimplifies complex human traits and behaviors. In summary, the concept of genetic determinism, while illuminating certain aspects of genetics, faces limitations in its ability to capture the intricate reality of human identity and behavior. The recognition of these limits encourages a more nuanced perspective that appreciates the complexity of gene-environment interactions and the enduring role of individual agency in shaping our lives. This philosophical exploration underscores that genetics is a multifaceted tapestry, and its influence is only one thread in the rich fabric of human existence.

The Uncertainty and Complexity of Genetic Research

As we journey deeper into the realm of genetic research, we are confronted with the profound notions of uncertainty and complexity. These aspects of genetic study challenge any simplistic or deterministic understanding of our genetic code and underscore the rich tapestry of intricacies within this scientific discipline. Uncertainty in genetic research emerges from the multifaceted nature of genetics itself. The genetic code is an intricate text, with each gene a paragraph, and each nucleotide a word in this cosmic script. Deciphering this code is a painstaking endeavor, and researchers continually unearth new layers of complexity. Geneticists encounter the challenges of genetic variations, including single nucleotide polymorphisms (SNPs) and structural variations, which introduce unpredictability and diversity into the genetic landscape. This diversity often leads to uncertainty when attempting to correlate specific genes with complex traits or diseases, as the genetic underpinnings are seldom straightforward. Complexity in genetic research arises from the interplay between genes, environmental factors, and their cumulative effects. It is not simply a matter of identifying individual genes that control a particular trait or disease; rather, it is a puzzle of how these genes interact with one another and with environmental influences. This intricate interplay gives rise to a web of complexity, as geneticists endeavor to untangle the multifaceted influences on our health, behaviors, and traits. The evolving understanding of epistasis, gene-environment interactions, and the influence of the microbiome all contribute to the complexity of genetic research. The recognition of uncertainty and complexity in genetic research has profound implications for our understanding of human health, identity, and the application of genetic knowledge. It challenges deterministic viewpoints, reminding us that the genetic code is not a crystal ball, but a dynamic, evolving script that leaves much to discover. Acknowledging these aspects underscores the need for humility in interpreting genetic data and the importance of embracing uncertainty as an inherent aspect of scientific exploration. In conclusion, the study of genetics unveils the complex and uncertain nature of our genetic code, serving as a reminder that our understanding of genes is a journey rather than a destination. Embracing this complexity and uncertainty enriches our appreciation of the intricate interplay between genes, environment, and individual agency, and reinforces the idea that genetics is a dynamic, evolving science that continually unravels the mysteries of life.

Bridging Genetics and Philosophy: Lessons for Both Fields

The intersection of genetics and philosophy offers profound insights for both disciplines, forging a dynamic bridge that enriches our understanding of the genetic world and the philosophical questions it provokes. This synergy between genetics and philosophy fosters a mutual exchange of wisdom, where each field contributes valuable lessons to the other. From the perspective of genetics, engaging with philosophy deepens the ethical and moral considerations inherent in genetic research. It compels geneticists to ponder the ethical implications of their work, encouraging them to consider the far-reaching consequences of genetic interventions and manipulations. This moral introspection has led to the development of ethical frameworks and guidelines that guide genetic research and the responsible application of genetic knowledge. The collaboration between genetics and philosophy serves as a reminder that science does not exist in isolation but must be grounded in ethical values and principles that respect human dignity and autonomy. Conversely, genetics also offers valuable insights to philosophy. The intricate study of genes sheds light on the complexities of identity, the nature of human traits, and the boundaries of determinism and free will. By exploring the genetic underpinnings of human behavior and biology, philosophy gains a deeper understanding of the intricate interplay between genes and environment and the role they play in shaping human existence. It challenges philosophical assumptions about human agency, autonomy, and the essence of human nature, inviting philosophical inquiries that enrich the discourse of both fields. Additionally, the interdisciplinary relationship between genetics and philosophy extends beyond ethics and metaphysics. It underscores the significance of interdisciplinary collaboration in scientific research. Geneticists and philosophers are increasingly partnering to address complex ethical dilemmas, engage in bioethical debates, and inform the public about the implications of genetic discoveries. This collaboration highlights the essential nature of a multidisciplinary approach to tackling the intricate challenges presented by genetic research, underscoring the interconnectedness of different fields in advancing human knowledge. In conclusion, the bridge between genetics and philosophy provides a fertile ground for mutual enrichment. Both fields learn valuable lessons from each other, deepening our understanding of the complex interplay between genetics, ethics, philosophy, and the broader human experience. This interdisciplinary partnership reminds us that science and philosophy are not solitary pursuits but rather interconnected disciplines that can together unravel the mysteries of life.

The Role of Ethics and Philosophy in Shaping Genetic Research

Ethics and philosophy play an integral role in shaping the trajectory of genetic research, offering valuable guidance and insights that transcend the boundaries of scientific inquiry. The partnership between genetics and ethics, along with the philosophical lens through which it is viewed, form a vital cornerstone for responsible and thoughtful exploration of the genetic realm. Ethics provides a moral compass for genetic research, ensuring that it aligns with values that respect human dignity, autonomy, and justice. As genetic capabilities advance, ethical considerations help in defining boundaries and offering guidance on issues such as informed consent, genetic privacy, and the responsible use of genetic technologies. Ethical frameworks and principles promote the equitable distribution of benefits and protections, reinforcing the importance of using genetic knowledge for the betterment of humanity. Through the ethical lens, genetic research is not merely a scientific endeavor but a moral journey that obliges researchers to navigate complex ethical terrain with responsibility and integrity. Philosophy, on the other hand, enriches genetic research by introducing broader questions about the implications of genetic discoveries for our understanding of the human condition. It delves into the philosophical dilemmas surrounding identity, free will, and determinism, as well as the ethical considerations related to individual autonomy and the collective responsibility of society. Philosophical inquiry encourages genetic researchers to explore the metaphysical dimensions of their work, pushing the boundaries of scientific inquiry and inviting contemplation about the profound ethical and philosophical questions that it raises. The collaboration between genetics, ethics, and philosophy underscores the importance of a multidisciplinary approach in genetic research. It serves as a reminder that science is not solely an empirical endeavor but a complex tapestry woven with ethical and philosophical threads. This interdisciplinary partnership fosters critical reflection, enriches the discourse, and guides genetic research along a path that respects the moral imperatives and philosophical intricacies inherent in the genetic quest for knowledge. In summary, the role of ethics and philosophy in shaping genetic research is fundamental. It ensures that genetic research is not pursued recklessly but is guided by ethical principles that safeguard human values and that it is accompanied by a philosophical exploration of the profound questions it poses. This dynamic partnership reinforces the idea that genetic research, in its pursuit of scientific advancement, must also be grounded in the ethical and philosophical foundations that inform the essence of human existence.

www.ingramcontent.com/pod-product-compliance
Lightning Source LLC
Chambersburg PA
CBHW072332290526
45794CB00002B/845